YOUR KNOWLEDGE HAS VALUE

Mohammad El-Goboushi, Amin Nassar

Improving Calic Compression Performance on both Continuous-Tone and Binary Images

GRIN Publishing

Bibliographic information published by the German National Library:

The German National Library lists this publication in the National Bibliography; detailed bibliographic data are available on the Internet at http://dnb.dnb.de .

Imprint:

Copyright © 2014 GRIN Verlag GmbH
Print and binding: Books on Demand GmbH, Norderstedt Germany
ISBN: 978-3-656-93073-0

This book at GRIN:

http://www.grin.com/en/e-book/294971/improving-calic-compression-performance-on-both-continuous-tone-and-binary

IMPROVING CALIC COMPRESSION PERFORMANCE ON BOTH CONTINUOUS-TONE AND BINARY IMAGES

Mohamed El-Ghoboushi*, Amin Nassar**

*Post graduate student, **Professor

Electronics and Electrical Communications Department

Cairo University, Faculty of Engineering, Giza, Egypt

mgobushi@yahoo.com

ملخص البحث:

ضغط الصور يعمل على تصغير أحجام الصور الرقمية مع الحفاظ على جودة الصورة. ويتحقق ذلك من خلال تطبيق أساليب ضغط البيانات. فعند تقليل حجم الصــــورة يكون هناك فقدان فى جودة الصورة أو لا. العمل المقدم هنا يبين ضغط الصور مع الحفاظ على جودتها باستخدام الترميز على أساس النمذجة بالقرائن الغير مؤثرة على جودة الصور (الكاليك) وقـد أظهرت النتائج أن ضغط الصور باستخدام الكاليك يوفر أعلى نسبة انضغاط من التقنيات الاخرى عند التجربة على الصور الطبيعية فى وقت قصير. فى هذا البحث نوضح كيف يمكن تعديل الكاليك للحصول على نسبة انضغاط أعلى مع الحفاظ على مساحة الذاكرة المستخدمة اذا كنا نتعامل مع الصور الطبيعية وكذلك تحسين نسبة الانظغاط بمتوسط 6,208 % فى حالة الصور الثنائية.

ABSTRACT

Context-based adaptive lossless image codec (CALIC) is one of the most efficient lossless encoding techniques for both continuous-tone and binary images. This Paper includes a research on how to modify CALIC algorithm in continuous-tone mode by truncating tails of the error histogram and using an escape mechanism to code the errors beyond the truncated code range, if they occur which improve CALIC compression performance. Also, we are going to propose a modification to CALIC in binary mode by eliminating error feedback mechanism. This minor modification should improve CALIC performance in binary images.

Keywords: *CALIC, Lossless Image Compression, Compression Ratio, Error Feedback, Binary mode, Continuous-tone Mode.*

1. INTRODUCTION

The assignment of compression is to code the image data into a compact form, minimizing both the number of bits in the representation, and the distortion caused by the compression.

There are two major approaches in the image compression field, which are lossless and lossy. A compression algorithm is "lossless" (or reversible) if the decompressed image is identical with the original. Respectively, a compression method is "lossy" (or irreversible) if the reconstructed image is only an approximation of the original one [2, 7]. The compression scheme presented in this Paper is a Lossless.

Predictive encoding is a major class of encoding scheme that is utilized in lossless compression. Context-based predictive [1] is a kind of adaptive predictive encoding, In which pixels are classified into different classes (a.k.m contexts) based on pixel neighborhood characteristics [8-14].

A suitable predictor for each context is adaptively selected and utilized for each context. As adaptive model changes the symbol probabilities during the compression process in order to adapt the statistics during the process [1]. Initially the compression process starts with an initial model, then during the process, the model adapts by the symbols, which have been transmitted already. Also, the decoder needs to be able to adapt the model in the same way later at the decompression process.

CALIC [1-3, 5-6][9-12] is an efficient context-based lossless image encoding technique. But it actually utilizes different size of entropy coding model to encode the prediction error ε based on different context error energies δ. Hence, even after error remapping, an alphabet size of 2^z is still unnecessarily large.

Large errors occur with diminishing frequency or not at all. But they still occupy spots in code space. If an error histogram of size 2^z is used, it will significantly reduce efficiency of entropy coding by distorting the underlying error statistics.

Hence, In this Paper we propose a modification by truncating the tails of the error histogram that is used to estimate *the probability p(e | δ)*, and using an escape mechanism to code the errors beyond the truncated code range, if they occur should improve CALIC performance in continuous-tone mode.

Also, CALIC attempts to solve binary image performance problem by incorporating a binary mode to deal with image regions with no more than two distinct grey levels. Unfortunately, when any pixel to be encoded has a different grey level than any of neighboring pixels, CALIC switches the algorithm to continuous-tone mode. This leads to downgrade the performance due to the error feedback step.

This paper is organized as follows: Section 2 gives an overview of CALIC algorithm and how it works, and discussing our proposed modifications of CALIC, Section 3 discusses our experimental work and results, respectively, and finally Section 4 offers our conclusion.

2. OVERVIEW OF CALIC

CALIC is a one pass context-based adaptive predictor scheme [2, 5]. CALIC operates in two modes: *binary* and *continuous tone modes*. The system selects one of the two modes on the fly during the coding process, depending on the context of the current pixel [9].

2.1 Continuous-tone mode

In continuous-tone mode, the neighborhood of the pixel to be encoded has more than two distinct grey levels. In this mode CALIC algorithm performs four operations [2, 5]:

a) Initial prediction,
b) Context classification [12],
c) Error feedback, and
d) Entropy encoding.

The initial prediction is obtained for the pixel to be encoded using Gradient Adjusted Predictor (GAP). GAP is a simple non-linear predictor that utilizes gradients at pixel neighborhood.

In context classification, each pixel is classified to one of the 576 predefined contexts. The context selection is based on comparing the value of the initial prediction with the pixel neighborhood's values.

For each context, CALIC assumes that the GAP predictor [2,4][5] is consistently repeating a similar prediction error. To compensate for this error, CALIC incorporates an error feedback stage [9], at which a bias value is added to the initial prediction. This bias value is the expectation of the prediction errors at the pixel's context.

Prediction errors are entropy encoded [11] using arithmetic encoder that utilizes context conditional probabilities, i.e., P(error|context) [9,15][16].

Unfortunately, CALIC utilizes different size of entropy coding model to encode the prediction error ε based on different context error energies δ. Hence, even after error remapping, an alphabet size of 2^z is still unnecessarily large.

Large errors occur with diminishing frequency or not at all. But they still occupy spots in code space. If an error histogram of size 2^z (where z is No. bits) is used, it will significantly reduce efficiency of entropy coding by distorting the underlying error statistics.

2.1.1 Proposed Modification of continuous-tone mode

Truncating the tails of the error histogram that is used to estimate the probability $p(e \mid \delta)$, and using an escape mechanism to code the errors beyond the truncated code range, if they occur should improve CALIC performance in continuous-tone mode.

- First we limit the size of each conditional error histogram to some value $N\delta$ based on the error energy level $1 \leq \delta < L$.
  Where, L is error energy level
- Second, For a given coding context δ, a symbol (remapped prediction error) $x \leq N\delta$ is encoded first to be $N\delta-1$, followed by the codeword for the new symbol $x - (N\delta-1)$, but being encoded under the coding context $\delta + 1$.
- The whole idea actually is similar to the post office will use different boxes to contain different parcels. And since we don't know the parcel size for time being, so we first

classify them by their estimated weight. Then if the parcel is oversize for the prescribed box of its estimate weight level, a larger box will be used.

2.2 Binary Mode

This mode is considered when a pixel's neighborhood has no more than two distinct grey levels [13]. It encodes the pixel's value directly from neighboring pixel values giving the code stream.

Unfortunately, when any pixel to be encoded has a different grey level than any of the neighboring pixels, CALIC triggers an escape sequence that switches the algorithm to continuous-tone mode. Hence, in this case the pixel will be treated as if it was in a continuous-tone region and then it will be encoded using GAP initial predictor and error feedback scheme.

2.2.1 Proposed Modification of binary mode

Using feedback mechanism in case of entering the continuous-tone mode as a result of an escape sequence triggered in binary mode may lead to decreasing the compression ratio.

Hence, we suppose to eliminate the positive feedback mechanism in this case only. This minor modification should improve CALIC performance in binary images. So we can use a *conditional feedback* in the overall system that allow using feedback mechanism in a *continuous-tone* mode and eliminate the feedback mechanism in case of entering the *continuous*-tone mode as a result of an escape sequence triggered in *binary* mode as shown in Fig. 1.

If the *binary* image has a different grey level than the two distinct levels of the neighboring pixels (i.e. T=2 where T is called symbolization.) the algorithm will eliminate feedback mechanism and code the error through the entropy coder directly. Otherwise, activate the continuous-tone mode with feedback mechanism.

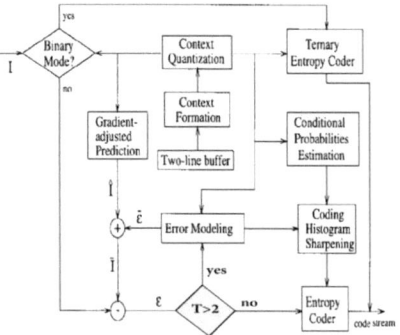

Fig. 1: The Overall modified CALIC system

3. EXPRIMENTS AND DISCUSSION

In this section the results of experiments are considered. For the experiments different test sets of images were used: natural "smooth" images and medical images.

Table 1. Test sets information.

Test set	Images	Average image size	Top number of colors	Bit-rate, bits per pixel
Medical images	5	328 x 290	256	8
Natural smooth Images	12	430 x 425	256	8

3.1 Modified Continuous-tone mode Results

Applying this proposed modification of CALIC algorithm on Lena image by truncating the error histogram with different thresholds according to different error

energies as shown in Fig. 2. hence, the symbols number for different entropy coding can be limited.

By doing a lot of experiments, CALIC got a good table sizes for the eight coding contexts as follows:

$$(N0 .... N7) = (18, 26, 34, 50, 66, 82, 114, 256) \quad (1)$$

We can check the thresholds and corresponding histogram in Fig. 2.

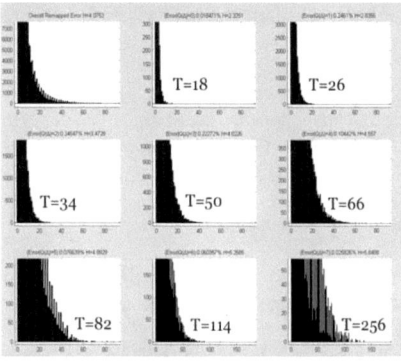

Fig.2. Histogram Truncation Threshold

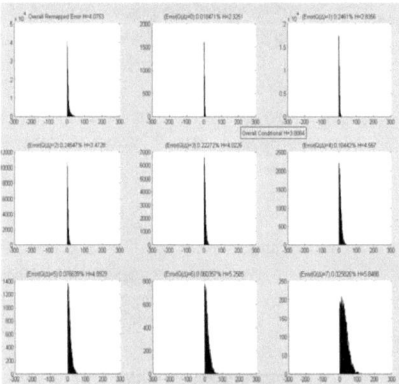

Fig.3. Error Histogram of original CALIC after Error Sign Flipping and Remapping.

After all of the above steps, here comes the final error for entropy coding.

We plot the error in Fig. 4 and we also marked the count of escape symbols introduced by histogram truncation.

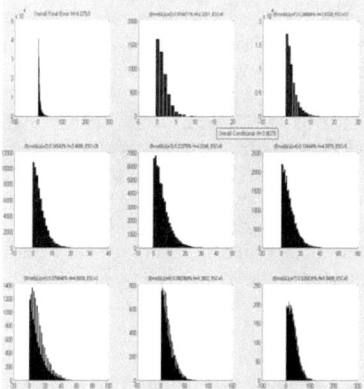

Fig.4. Error for Entropy Coding

- Comparing each plot in Fig.3 within Fig.2 we found the **following results:**

Table 2: Comparison results of the modified CALIC and the original.

Plot no.	No. of symbols in original CALIC	No. of bits needed	No. of symbols in modified CALIC	No. of bits needed
(1)	90	7	75	7
(2)	18	5	18	5
(3)	39	6	26	5
(4)	70	7	34	6
(5)	59	6	50	6
(6)	71	7	66	7
(7)	85	7	82	7
(8)	90	7	90	7
(9)	140	8	140	8

It can be seen that the results on continuous-tone images, modified CALIC really works well as it saves at most one bit and then the code execution will be faster and takes less time.

Results are given in bit-rate and evaluated in bits per pixel. Initial bit-rate for all images is **8 bits per pixel.**

Table 3. Bit-rate of compression results on natural smooth images (bits per pixel).

File name	Modified CALIC	JPEG2000	LOCO-I
Lena	4.2055	4.313	4.237
Peppers	3.7743	4.625	4.489
Barb	4.9143	4.781	4.863
Boat	4.8350	4.406	4.250
C man	3.8729	5.000	4.740
Flowers	5.7865	4.021	3.914
Einstein	3.9017	4.813	4.602
Goldhill	3.6105	4.844	4.712
Zelda	3.2509	4.000	4.005
Man	4.5580	5.500	5.267
Elaine	5.0110	4.938	4.898
Mri	4.0276	4.600	4.518
Average	**4.3123**	4.653	4.541

3.2 Modified Binary mode results

Simulating CALIC algorithm without feedback mechanism on images that have no more than two values or also, may have a different grey level of neighboring pixels like medical images, Fig.4, the results are summarized in table 4.

Table4. Bit-rate of compression results on medical images (bits per pixel) after eliminating feedback mechanism compared to JPEG2000, LOCO-I and CALIC with feedback mechanism.

File name	CALIC without feedback	JPEG 2000	CALIC with feedback	LOCO-I
Blood	3.8289	4.682	4.0419	4.076
Hip	2.3889	2.611	2.6019	1.887
Ultrasou	3.1216	4.886	3.3346	4.679
5week	3.5902	3.576	3.8038	3.325
Child	3.1602	3.83	3.3732	3.683
Average	**3.2179**	3.917	3.4309	3.530

From table 4 we find that the proposed modified CALIC algorithm improved the compression performance over the original CALIC in all binary test images. On average, the compression performance has been improved by 6.208% on binary images. This supports our claim that the error feedback is not a good technique to be used on regions of the image containing few widely separated grey levels.

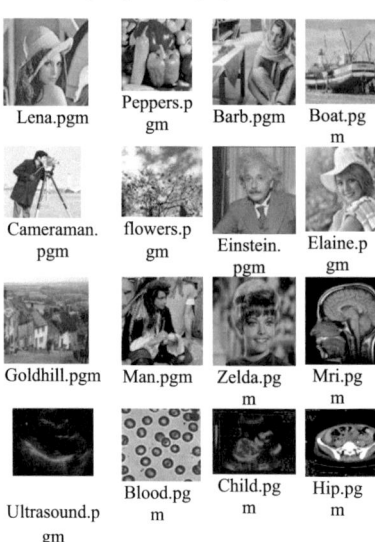

Lena.pgm Peppers.pgm Barb.pgm Boat.pgm

Cameraman.pgm flowers.pgm Einstein.pgm Elaine.pgm

Goldhill.pgm Man.pgm Zelda.pgm Mri.pgm

Ultrasound.pgm Blood.pgm Child.pgm Hip.pgm

5week.pgm

Fig.4: Different test sets of images were used: natural "smooth" images and medical images

4. Conclusion

As CALIC actually utilizes different size of entropy coding model to encode the prediction error ε based on different context error energies δ, an alphabet size of 2^z is considered unnecessarily large.

We proposed truncating the tails of the error histogram on continuous-tone images. Our modification improved the compression performance of CALIC on continuous-tone mode as it saves at most one bit and then the code execution will be faster and takes less time.

Also, CALIC compression ratio is decreased on images with fewer and widely separated grey levels (i.e. binary images). This is due to the error feedback mechanism which is considered in the case of entering the continuous-tone mode as a result of an escape sequence triggered in binary mode.

We proposed using a conditional error feedback mechanism, where the algorithm will eliminate the feedback in binary regions and perform it only on continuous-tone regions of the image. Our modification improved the compression ratio of CALIC on binary images by 6.208% (on average).

ACKNOWLDGEMENT

Most importantly, We are ever grateful to **GOD** the major source of strength when we worked on this paper and it is under His grace that we live and learn.

REFERENCES

[1] Grzegorz Ulacha, and Ryszard Stasiński, "*Effective Context Lossless Image Coding Approach Based on Adaptive Prediction*", World Academy of Science, Engineering and Technology, September 2009, Vol.3, pp. 63-68.

[2] Arpita C. Raut, Dr. R. R. Sedamkar, Mumbai University, "*Dynamic Efficient Prediction Approach for Lossless Image Compression*", International Journal of Advancements in Research & Technology, September 2013, Volume 2, Issue 9, pp. 132 – 137.

[3] Hengjian Li, Lianhai Wang, Song Zhao, Zutao Zhang, "*Research on*

Lossless Compression Algorithms of Low Resolution Palmprint Images", Research Journal of Applied Sciences, Engineering and Technology 4(14), July 2012, pp. 2072-2081.

[4] Anil Mishra, Vinamra Khand, Gomti Nagar, Lucknow, "A PREDICTIVE CODING METHOD FOR LOSSLESS COMPRESSION OF IMAGES", INTERNATIONAL JOURNAL OF COMPUTERS & TECHNOLOGY, June 2013, Vol 7, No 3, pp. 683 – 685.

[5] AmrutaShinkar, Dr. S. A. Patekar, Prof. Mandar Sohani, "ADAPTIVE LOSSLESS IMAGE COMPRESSION USING SUPER SPATIAL STRUCTURE PREDICTION", International Journal of Advanced Research in Computer Science and Electronics Engineering (IJARCSEE) , November 2013, Volume 2, Issue 11, pp. 718 – 721.

[6] Hengjian Li, Lianhai Wang, Song Zhao, Heng Yang, " A Comparison of Lossless Compression Methods for Palmprint Images", JOURNAL OF SOFTWARE, March 2012, VOL. 7, NO. 3, pp. 594 – 598.

[7] P. Fränti, "Image Compression", University of Joensuu, Department of Computer Science, Lecture notes, 2000.

[8] M. Weinberger, G. Seroussi, and G. Sapiro, "LOCO-I: a low complexity, context-based, lossless image compression algorithm", Proceedings of Data Compression Conference, March 1996, pp. 140 – 149.

[9] X. Wu and N. Memon, "Context-based, adaptive, lossless image coding", IEEE Transactions on communications, April 1997, Vol. 45, No. 4, pp. 437 – 444.

[10] X. Wu, "Lossless compression of continuous-tone images via context selection, quantization, and modeling" , IEEE Transactions on Image Processing, May 1997, Vol. 6, No. 5, pp. 656 – 664.

[11] N. Memon and X. Wu, "Recent Developments in Context Based Predictive Techniques for Lossless Image Compression", Computer Journal, November 1997, Vol. 40, No. 2 and 3, pp. 127 – 136.

[12] F. Golchin and K. Paliwal, "A lossless image coder with context classification, adaptive prediction and adaptive entropy coding", Proceedings of IEEE International Conference on Acoustics, Speech, and Signal Processing, May 1998, Vol. 5, pp. 2545 – 2548.

[13] X. Wu and N. Memon, "Context-based lossless interband compression-extending CALIC", IEEE Transactions on Image Processing, June 2000, Vol. 9, No. 6, pp. 994 – 1001.

[14] M. Weinberger, G. Seroussi, and G. Sapiro, "The LOCO-I lossless image compression algorithm: principles and standardization into JPEG-LS", IEEE Transactions on Image Processing, August 2000 , Vol. 9, No. 8, pp. 1309 – 1324.

[15] X. Wu, "CALIC – A Context-Based, Adaptive, Lossless Image Codec", IEEE Trans. Communications, April 1997, vol. 45, No. 4.

[16] Mehmet Utku Celik, Gaurav Sharma, A. Murat Tekalp, "Gray-level-embedded lossless image compression", Elsevier Science B.V. on Signal Processing: Image Communication, January 2003, pp. 443 – 454.